Superheated Steam Drying

A Smart and Sustainable Approach to Food Preservation

Published 2026 by River Publishers
River Publishers
Broagervej 10, 9260 Gistrup, Denmark
www.riverpublishers.com

Distributed exclusively by Routledge
605 Third Avenue, New York, NY 10017, USA
4 Park Square, Milton Park, Abingdon, Oxon OX14 4RN

Superheated Steam Drying: A Smart and Sustainable Approach to Food Preservation / Dipali Saxena and A. K. Haghi.

Routledge is an imprint of the Taylor & Francis Group, an informa business

ISBN 978-87-438-1018-6 (paperback)
ISBN 978-87-438-1020-9 (online)
ISBN 978-8-743-81019-3 (ebook master)
ISBN 978-87-438-1022-3 (ePub)

DOI 10.1201/9788743810193

Superheated Steam Drying

A Smart and Sustainable Approach to Food Preservation

Dipali Saxena
Shri Vaishnav Vidyapeeth Vishwavidyalaya,
India

A. K. Haghi
University of Coimbra,
Portugal

River Publishers

Routledge
Taylor & Francis Group
NEW YORK AND LONDON

Contents

Preface

As the global food industry increasingly embraces sustainability and intelligent processing, there is a critical need for technologies that can reduce environmental impact while preserving the quality, safety, and nutritional value of food. Among these innovations, *superheated steam drying (SHS)* has emerged as a transformative technique, offering oxygen-free, energy-efficient drying that minimizes oxidative degradation and enhances product quality. This book, *Superheated Steam Drying: A Smart and Sustainable Approach to Food Preservation*, presents a comprehensive exploration of SHS technology, highlighting its scientific principles, engineering components, and broad applications across food and biological materials. It discusses the advantages of SHS over conventional drying methods and emphasizes its potential to meet both industrial and environmental demands. The book also focuses on hybrid drying systems that combine SHS with microwave, infrared, and vacuum technologies to deliver customized and highly efficient drying solutions. In line with modern digital advancements, it includes detailed insights into the integration of automation, IoT, and smart control systems, demonstrating how SHS drying can evolve into a responsive, self-regulating process aligned with Industry 4.0 goals. Furthermore, it addresses the policy frameworks and institutional support necessary for scaling SHS technologies in global food systems. Designed for food technologists, engineers, researchers, environmental specialists, and policymakers, this book serves as a valuable resource for advancing sustainable food preservation. We extend our sincere thanks to all contributors and collaborators who have helped shape this work, and we hope it inspires continued innovation and adoption of SHS as a cornerstone of smart, eco-conscious food processing.

About the Authors

Dr. Dipali Saxena is an accomplished Assistant Professor with over a decade of academic and research experience in Food Science and Technology. She earned her Ph.D. in Food Technology from Jiwaji University, Gwalior India, specializing in Nutraceuticals and Functional Foods, and graduated as the University Topper and Gold Medalist in M.Sc. Food Technology. Currently, she serves as Assistant Professor in the Department of Food Nutrition at Shri Vaishnav Institute of Home Science, under Shri Vaishnav Vidyapeeth Vishwavidyalaya, Indore. A passionate educator and researcher, Dr. Saxena mentors Ph.D. and M.Sc. scholars and has designed vocational and elective courses in alignment with India's National Education Policy 2020. Dr. Saxena has authored over 45 research publications in national and international journals and conferences. Dr. Saxena continues to contribute to advancing food technology education and sustainable food innovation through her research on non-thermal preservation, functional foods, and nutraceuticals.

A. K. Haghi is a retired professor and is currently a research associate at the University of Coimbra, Portugal. Professor Haghi holds a B.Sc. in urban and environmental engineering from the University of North Carolina (USA) and holds two M.Sc. degrees, one in mechanical engineering from North Carolina State University (USA) and the other in applied mechanics, acoustics, and materials from the Université de Technologie de Compiègne (France). He was awarded a Ph.D. in engineering sciences at Université de Franche-Comté (France). He has written, co-written, edited, or co-edited more than 1000 publications, including books, book chapters, and papers in refereed journals with over 4300 citations and an h-index of 35, according to the Google Scholar database.

CHAPTER

1

Introduction to Superheated Steam Drying Technology

Drying is one of the oldest food methods, dating back to ancient civilizations where sun and wind drying were used to extend the shelf life of perishable goods. Traditional air drying was the foundation of dehydration processes and was extensively used in agriculture, textiles, and ceramics. With the advent of the Industrial Revolution, drying technologies became more mechanized, leading to the development of convection, conduction, and radiation-based dryers (Mujumdar, 2014). Early industrial drying relied heavily on hot air systems, due to their simplicity and low initial cost. However, such methods often resulted in high energy consumption, inconsistent product quality, and environmental concerns due to high carbon emissions.

In the late 20th and early 21st centuries, increased awareness of energy conservation and sustainability triggered innovation in thermal drying technologies. With drying processes accounting for up to 15% of industrial energy use globally, researchers began exploring more efficient and environmentally friendly alternatives. Superheated steam drying (SSD) emerged as a promising technique offering multiple benefits such as recovery and reuse of latent heat, oxygen-free environment reducing oxidation, potential integration with waste heat recovery systems, and reduction in specific energy consumption by 20–50% compared to hot air drying (Jangam and Mujumdar, 2010). SSD's development was also influenced by stricter environmental regulations and rising fuel prices, which made energy efficiency and carbon footprint reduction key industrial goals.

DOI: 10.1201/9788743810193-1

1.1 Conventional Drying Methods and Superheated Steam Drying (SSD)

Drying is a crucial unit operation in food processing, aimed at extending shelf life by reducing moisture content and inhibiting microbial growth. Traditional drying methods primarily involve the use of hot air, direct sunlight, infrared radiation, or contact heat to evaporate water from food materials.

Hot air drying (HAD) is the most widely used method in the food industry. In this technique, heated air is passed over the food to evaporate moisture. Although effective, HAD often results in oxidation of nutrients, color degradation, shrinkage, and longer drying times. It is also energy-intensive and has limited energy recovery potential. Open sun drying is one of the oldest drying methods, predominantly used in rural areas and small-scale food processing. It is weather-dependent, slow, and unhygienic due to exposure to dust, insects, and animals. Solar drying is an improved version that uses solar collectors and enclosures to enhance drying efficiency. Vacuum drying and freeze-drying are advanced techniques used for high-value, heat-sensitive foods. These methods can preserve product quality effectively but are expensive and require complex equipment and high energy input.

1.2 Limitations of Conventional Methods

Conventional drying methods in the food industry face several significant limitations. They often involve high energy consumption and operational costs, making them economically challenging, especially for large-scale applications. Additionally, the use of high temperatures can lead to quality deterioration in food products due to oxidation and nutrient degradation. These methods also exhibit inefficient energy recovery, further contributing to energy wastage. Moreover, they pose environmental concerns, as they can result in the emission of volatile organic compounds (VOCs) and greenhouse gases, thereby impacting air quality and contributing to climate change.

1.3 Introduction to Superheated Steam (SHS)

Superheated steam (SHS) is defined as steam that has been heated beyond its saturation temperature at a given pressure, meaning it contains no moisturize droplets and behaves like a dry gas. Unlike saturated steam, which is in equilibrium with water at a specific pressure and temperature, superheated steam does not condense when it comes in contact with cooler surfaces—unless it is cooled below its saturation temperature. In drying applications, superheated steam acts as a drying medium by absorbing moisture from wet solids and carrying it away in vapor form. Since SHS contains more energy than saturated steam, it can be efficiently used for thermal drying without introducing oxygen, which significantly improves the drying environment for heat-sensitive or oxidation-prone materials like food.

Superheated steam is defined as steam (Figure 1.1) that has been heated beyond its saturation (boiling) point at a given pressure, without containing any liquid water droplets. In this state, it behaves more like dry gas than a mist and can be used as a drying medium in place of hot air.

In thermodynamic terms, superheated steam exists in the region beyond the saturated steam line on a pressure-enthalpy or temperature-entropy diagram. It possesses more energy than saturated steam and does not condense immediately upon contact with a surface, making it an effective heat transfer medium.

1.3.1 Characteristics of superheated steam

Superheated steam is steam that has been heated beyond its boiling point at a given pressure, such as above 100 °C at atmospheric pressure, and exists entirely in a gaseous phase without any moisture content. One of its key characteristics is the absence of oxygen, which creates a non-oxidizing environment ideal for preserving sensitive food components. This oxygen-free condition prevents oxidation-related spoilage, thus protecting the color, flavor, texture, and nutritional value of food during the drying process. Additionally, superheated steam possesses high thermal energy, enabling faster and more efficient heat transfer to the product being dried.

The use of superheated steam in drying applications presents numerous advantages over conventional hot air drying. Its non-oxidizing nature significantly reduces the risk of degradation in food quality, making it particularly suitable for heat-sensitive or high-value products. Moreover, the energy utilized in the drying process can be partially recovered by condensing the used steam, thereby improving overall energy efficiency. This capability not only reduces operational costs but also supports sustainability goals. The rapid drying rates associated with superheated steam contribute to shorter processing times and enhanced throughput. Furthermore, since it minimizes the release of volatile organic compounds (VOCs) and facilitates the recovery of waste heat, the technology is environmentally friendly and aligns with modern energy conservation and emission reduction strategies.

Figure 1.1 Formation of steam.

1.3.2 Evolution of SSD

The idea of using steam for drying is not new. In the early 20th century, steam-based drying was explored in the pulp and paper industry, where steam cylinders were used to dry paper sheets. However, these systems used saturated steam and were not considered SSD systems.

The concept of superheated steam as a drying medium began gaining interest during the mid-20th century (1950s–1970s). Initial research focused on the thermodynamic advantages of steam and its ability to be reused in closed-loop systems. However, technical limitations such as difficulty in controlling temperature and pressure, equipment costs, and a lack of understanding of heat and mass transfer mechanisms restricted its application.

SSD started being studied systematically in the 1980s with the development of mathematical models for heat and mass transfer under SHS conditions. Pioneering contributions from researchers such as Professor Arun S. Mujumdar and Dr. S.V. Jangam laid the foundation for this field through their extensive publications and experimental work. During this period, laboratory and pilot-scale SSD units were developed, and the technology was tested primarily for drying heat-sensitive food products such as vegetables, fruits, and meats. These early studies significantly advanced the understanding of drying kinetics and highlighted the importance of implementing closed-loop control systems to optimize efficiency and product quality. From the 2000s onward,

SSD has experienced growing interest and commercial application, driven by global efforts to conserve energy and reduce carbon emissions. Technological advancements in materials science, process automation, and computational modeling have expanded SSD's applicability beyond food processing to sectors such as pharmaceuticals, where it is used to dry heat-sensitive active compounds, and environmental engineering, where it is applied to drying materials like sewage sludge and biomass. In the food industry, SSD has proven effective for products such as herbs, spices, fish, and dairy powders, making it a versatile and environmentally sustainable drying solution across multiple domains.

1.3.3 Sources and generation of SHS

Superheated steam is usually produced through a two-step process. First, saturated steam is generated using a conventional boiler by boiling water at a specific pressure. This saturated steam contains moisture and is at the boiling point for that pressure. To convert it into superheated steam, the saturated steam is then passed through a device called a *superheater*. The superheater increases the temperature of the steam beyond its boiling point without significantly increasing its pressure. As a result, the steam becomes completely dry, with no water droplets, and reaches a much higher temperature.

The exact temperature of the superheated steam depends on several factors, including the type of product being dried, how sensitive it is to heat, and the overall design of the drying system. In most modern applications, superheated steam is used in a closed-loop system. This means that after the steam has been used for drying, it is collected and separated from the moisture that was removed from the product. The clean steam is then reheated using the superheater and reused in the system. This method not only conserves energy but also helps in maintaining a consistent drying environment, making the process more efficient and environmentally friendly.

1.4 Principles of Superheated Steam Drying

Superheated steam drying (SSD) is an advanced drying technique that utilizes steam heated beyond its saturation temperature to remove moisture from wet materials. The process begins with the generation of saturated steam, which is then heated further using superheaters to convert it into superheated steam. This steam, having a higher temperature and enthalpy than saturated steam, becomes a powerful medium for heat transfer. When it comes into contact with the moist material—either by passing over, through, or around it—the high thermal energy causes rapid evaporation of surface moisture. As the drying continues, internal moisture migrates to the surface and evaporates, gradually reducing the overall moisture content of the material.

As moisture evaporates from the material, it combines with the superheated steam atmosphere inside the drying chamber. This vapor-laden steam is then carried out of the chamber. The moist steam can either be condensed, with latent heat recovered for reuse, or reheated and recycled into the drying system. This makes SSD a closed-loop system, enhancing energy efficiency and minimizing thermal losses. The ability to recover and reuse energy, especially the latent heat of vaporization, is one of the key benefits of this method. This results in a significant reduction in overall energy consumption compared to traditional hot air drying methods.

The efficiency and effectiveness of SSD are largely due to the favorable thermodynamic properties of superheated steam. It has high enthalpy, which ensures rapid heat transfer, and lower density compared to liquid water, allowing it to flow easily through and around the material being dried. Moreover, superheated steam acts as an inert medium, as it contains no oxygen. This greatly reduces the risk of oxidation during drying, making SSD particularly suitable for materials that are sensitive to oxidative degradation, such as food products, pharmaceuticals, and certain chemicals.

1.5 Advantages of Superheated Steam Drying Over Conventional Drying Methods

Drying processes are essential in the food, pharmaceutical, chemical, and textile industries. Conventional drying methods—particularly hot air drying (HAD)—are commonly used due to their simplicity and familiarity. However, these methods are increasingly being replaced by modern techniques like superheated steam drying (SSD), owing to their superior efficiency, product quality retention, and environmental sustainability. Below is a comprehensive discussion on the major advantages of SSD over conventional drying (Figure 1.2 and Table 1.1).

1.5.1 Improved product quality

One of the most significant advantages of SSD is its ability to preserve the physical, nutritional, and sensory characteristics of the product. In hot air drying, the presence of oxygen at elevated temperatures leads to oxidation of sensitive compounds, such as vitamins, lipids, pigments, and proteins. This results in loss of color, flavor, aroma, and nutritional value (Ratti, 2001).

Figure: 1.2 Advantages of superheated steam drying technology.

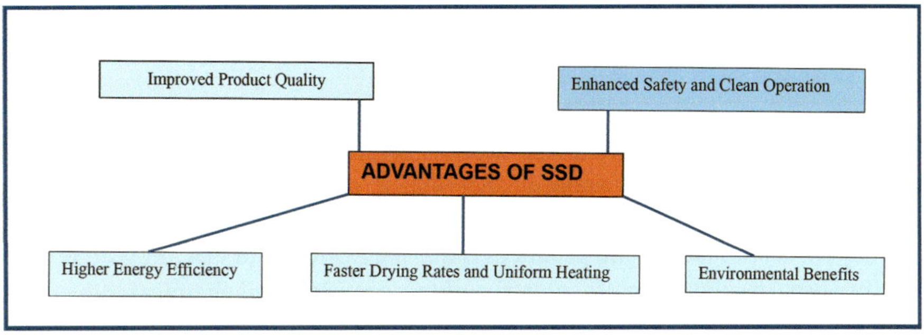

Table 1.1 Advantages of SSD vs. conventional drying.

Parameter	Conventional drying (hot air)	Superheated steam drying (SSD)
Medium	Air (with oxygen)	Steam (oxygen-free)
Oxidation risk	High	Low
Product quality	Moderate to poor	High (color, nutrients, flavor retained)
Energy consumption	High	Lower (with energy recovery)
Environmental impact	High emissions	Low emissions, recyclable medium
Fire/explosion risk	Present	Minimal
Drying rate	Moderate	Faster and more uniform

1.5.2 Higher energy efficiency and recovery

In traditional drying systems, much of the energy used for heating air is lost through exhaust gases. SSD allows for better heat utilization because steam can be collected, dehumidified, and reused, minimizing energy losses. The condensation of steam after moisture absorption can be used to recover latent heat (Mujumdar, 2014). This closed-loop system significantly reduces the overall energy consumption compared to open-loop hot air drying. Studies show that SSD can achieve 10–25% energy savings, depending on the type of material and process design (Bade et al., 2024).

1.5.3 Faster drying rates and uniform heating

Superheated steam, due to its higher thermal conductivity and energy content compared to air, enables faster heat transfer to the material being dried. This results in shorter drying times, more uniform heating, and less overheating of surface layers, minimizing crust formation and hardening. For example, the drying time for thin slices of carrots and apples was reduced by nearly 30% using SSD compared to HAD (Devahastin, 2007).

1.5.4 Enhanced safety and clean operation

Hot air dryers operate in the presence of oxygen, which increases the risk of fire and explosion, especially when drying volatile or combustible materials. SSD offers a safer environment because it operates in a non-oxidizing medium, does not support combustion, and produces fewer particulate and VOC emissions. This feature is particularly beneficial in the pharmaceutical and chemical industries, where explosion-proof environments are necessary (Strumiłło and Kudra, 1986).

1.5.5 Environmental benefits

SSD offers considerable environmental advantages such as reduced emissions of volatile organic compounds (VOCs) and odors due to the closed-loop system, and a lower carbon footprint as energy consumption is reduced. Recyclability of steam leads to minimal heat and moisture waste.

In wastewater treatment and sludge drying, SSD has proven effective in reducing greenhouse gas emissions and offensive odors compared to conventional thermal dryers (Kudra and Mujumdar, 2009).

1.5.6 Possibility of combined heat and power (CHP) integration

SSD systems can be coupled with combined heat and power (CHP) setups where the steam generated by waste heat recovery systems is utilized for drying. This integration further enhances energy efficiency and sustainability in industrial operations (Bade et al., 2024).

References

Bade, M.H., Jangam, S.V., & Mujumdar, A.S. (2024). *Superheated Steam Drying: Technology for Improved Sustainability and Quality*. CRC Press.

Devahastin, S. (2007). *Food and Bioproducts Processing*, 85(2), 109–114.

Kudra, T., & Mujumdar, A.S. (2009). *Advanced Drying Technologies*. CRC Press.

Mujumdar, A.S. (2014). *Handbook of Industrial Drying* (4th ed.). CRC Press.

Strumiłło, C., & Kudra, T. (1986). *Drying: Principles, Applications and Design*. Gordon and Breach.

CHAPTER

2

Principles of Superheated Steam Drying

Superheated steam drying (SSD) is a progressive, energy-efficient, and environmentally sustainable technique that has emerged as a viable alternative to conventional hot air drying. It involves the use of steam that is heated beyond its saturation point and is devoid of moisture, enabling it to act as a superior drying medium. Unlike air-drying systems, SSD operates in an oxygen-free environment, significantly reducing the risk of oxidation, enhancing product quality, and improving energy efficiency.

This chapter elaborates on the fundamental principles governing SSD, focusing on its heat and mass transfer mechanisms, energy efficiency, energy-saving characteristics, and environmental benefits and challenges.

2.1 Heat and Mass Transfer Mechanisms

In any drying process, the fundamental physical phenomena governing the removal of moisture from materials are heat transfer and mass transfer. These two mechanisms operate simultaneously and are highly interdependent. In the case of superheated steam drying (SSD), the dynamics of heat and mass transfer differ significantly from those in conventional air-drying systems due to the unique thermophysical properties of superheated steam (Figure 2.1).

Superheated steam, by definition, is steam that has been heated beyond its saturation temperature at a given pressure. This state imparts it with several beneficial characteristics, including higher thermal conductivity and

DOI: 10.1201/9788743810193-2

specific heat capacity when compared to ambient air. These attributes allow superheated steam to serve as an efficient medium for heat transfer during the drying process. One of the most important features of superheated steam is its ability to condense on the surface of the drying material. When condensation occurs, it releases a substantial amount of latent heat, thereby accelerating the heating of the product's surface and enhancing the drying rate.

Figure 2.1 Principle of superheated steam drying.

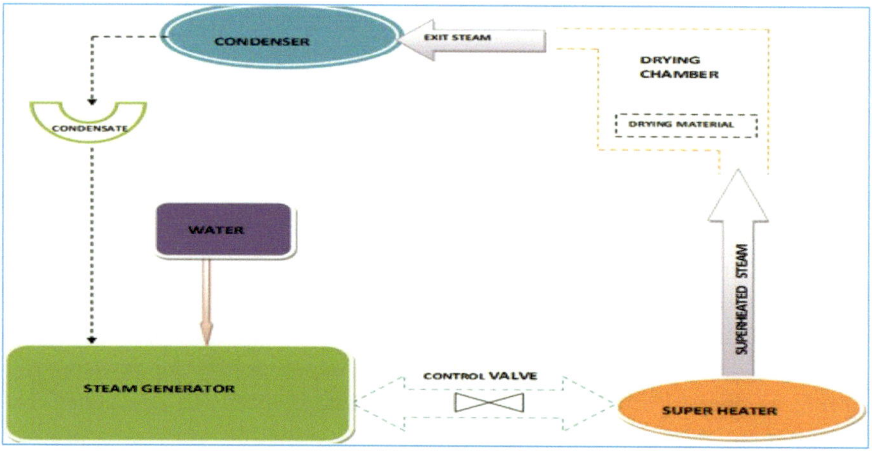

Heat transfer in SSD occurs primarily through forced convection, where superheated steam flows at high velocity over or through the moist material. Upon contact with the relatively cooler surface of the wet material, the steam may lose a portion of its superheat, leading to the transfer of thermal energy to the product. This energy transfer serves multiple functions: it raises the surface temperature of the material, initiates the internal migration of moisture due to a thermal gradient, and promotes surface evaporation of water molecules. The effectiveness of this heat transfer process is amplified when the steam is maintained at a temperature significantly higher than the boiling point of water (i.e., above 100 °C at atmospheric pressure, or even higher under pressurized conditions), ensuring a strong temperature gradient and thus more rapid drying.

The mass transfer mechanism in SSD is equally crucial to the overall drying efficiency. Moisture contained within the material must move from the interior to the surface before it can be evaporated. This is governed primarily by internal diffusion, which is driven by the vapor pressure gradient established between the moist interior and the drier surface of the material. As the temperature of the product increases, the vapor pressure inside the material rises, facilitating the migration of water vapor to the surface. Once the moisture reaches the surface, it is evaporated due to the high external temperature and the inherently low relative humidity of superheated steam, which acts as a drying medium.

The third stage of mass transfer involves the removal of vaporized moisture from the vicinity of the material into the surrounding steam atmosphere. Since superheated steam has a lower partial pressure of water vapor compared to saturated steam or humid air, it creates favorable conditions for continuous evaporation. Furthermore, because the steam is in a superheated state, it does not readily become saturated, allowing high evaporation rates to be sustained over time without the need for dehumidification—an issue common in traditional air-drying systems.

2.2 Energy Efficiency Considerations

Energy efficiency is a cornerstone of superheated steam drying (SSD) technology, offering a significant advantage over traditional hot air drying methods (Kumar, 2023). Conventional dryers often face substantial energy losses due to the simultaneous heating of both air and moisture, the continuous expulsion of warm, moist air, and the loss of latent heat that cannot be recovered. In contrast, SSD systems utilize steam as both the heating medium and the drying atmosphere. During drying, the moisture evaporated from the product merges with the steam stream and is subsequently recovered through condensation or mechanical vapor recompression (MVR), then re-superheated and reused within a closed-loop system. This recycling minimizes energy losses and significantly enhances thermal efficiency. As noted by Mujumdar (2014), SSD systems can achieve up to 40% higher energy efficiency compared to traditional air dryers when equipped with advanced heat recovery and MVR technologies. Moreover, SSD can be operated under slightly pressurized conditions, which elevates the boiling point of water and enables more precise temperature control—contributing to further energy savings and better compatibility with heat-sensitive materials.

2.3 Energy Saving Characteristics

The energy-saving potential of SSD stems from its closed-loop configuration, reduced thermal losses, and integration capability with heat recovery and cogeneration systems. A study by Lee and Kwon (2022) showed up to 30%

energy reduction in SSD processes for tofu skin compared to hot air drying, attributed mainly to reduced process time and heat recycling.

Key aspects of energy saving in SSD:

(i) Steam condensation recovery: Condensed steam (from evaporated product moisture) retains latent heat, which can be reused in the process, either by direct reuse or through steam re-superheating systems.
(ii) Mechanical vapor recompression (MVR): MVR compresses and reheats low-pressure steam, allowing it to be recycled within the dryer at minimal additional energy cost (Bade et al., 2024).
(iii) Shorter drying times: Due to superior heat transfer properties, SSD can reduce drying times by 20–50%, thereby reducing total energy demand.
(iv) No need for exhaust air dehumidification: In hot air systems, air must be constantly replaced or dried; SSD avoids this, as the medium (steam) remains dry by design.

2.4 Environmental Benefits and Challenges

From an environmental standpoint, SSD offers numerous benefits. The absence of oxygen prevents oxidative deterioration and eliminates emissions of volatile organic compounds (VOCs) and odors. Its closed-loop operation minimizes the release of moisture and heat into the environment, leading to lower CO_2 and NO_x emissions compared to traditional fossil-fuel-based hot air dryers, especially when powered by renewable or waste heat sources. Additionally, SSD improves workplace hygiene and safety by eliminating airborne dust and particulates; condensed steam can also remove potential allergens or pathogens, as noted by Kudra and Mujumdar (2009). SSD is particularly advantageous in applications involving biosolids and waste materials, where it significantly aids in odor control and reduces the reliance on chemical treatments.

2.5 Conclusion

Superheated steam drying is a multifaceted, high-efficiency thermal process that enhances heat and mass transfer, optimizes energy usage, and supports green manufacturing. By combining thermodynamic precision with closed-loop efficiency, SSD emerges as a leading choice for industries seeking energy savings and environmental sustainability. While it involves higher initial investments and technological complexity, the long-term benefits of product quality retention, reduced carbon emissions, and energy recovery outweigh its limitations, paving the way for broader industrial adoption.

References

Bade, M. H., Jangam, S. V., & Mujumdar, A. S. (2024). *Superheated Steam Drying: Technology for Improved Sustainability and Quality*. CRC Press.

Lee, S., & Kwon, T. H. (2022). Energy and product quality improvements in tofu skin drying using superheated steam drying. *Journal of Food Engineering*, 321, 110974. https://doi.org/10.1016/j.jfoodeng.2022.110974

Kumar, A., Shukla, S., & Singh, S. (2023). Advances in superheated steam drying and its application in food processing. *Drying Technology*, 41(12), 1345–1362. https://doi.org/10.1080/07373937.2023.217528

CHAPTER

3

Elements of Superheated Steam Drying Technology

Superheated steam drying (SSD) is an emerging thermal drying technique increasingly utilized in the food, pharmaceutical, and biomass industries. It involves the use of steam heated above its saturation temperature, enabling it to function as a highly effective and inert drying medium. Compared to traditional hot air drying, SSD offers several distinct advantages such as higher drying efficiency, oxygen-free processing, and reduced energy loss. These benefits are particularly significant in food applications where maintaining product quality, nutritional value, and color is critical (Mujumdar, 2014). Furthermore, SSD systems often operate in closed-loop configurations, which allow for the recovery of latent heat and water, improving overall sustainability and lowering environmental impact.

3.1 Overview of Superheated Steam Drying

In SSD, the drying medium—superheated steam—has a temperature higher than the boiling point of water at a given pressure and does not contain water droplets. When it comes into contact with a moist material, heat transfer primarily occurs through convection, rapidly heating the material surface and causing surface moisture to evaporate. This evaporated water is swept away by the steam flow, creating a concentration gradient that drives internal moisture to the surface. One of the major advantages of SSD is its operation in an oxygen-free environment, which prevents oxidative reactions, browning, or nutrient degradation, thus preserving food quality. The non-flammable and recyclable

DOI: 10.1201/9788743810193-3

nature of the steam medium also improves safety and makes it suitable for sensitive products.

3.2 Types of Superheated Steam Dryers Used in the Food Industry

Superheated steam dryers used in the food industry can be broadly categorized into direct contact and indirect contact systems, each designed to meet specific processing needs based on the nature of the food material and desired product characteristics. The choice between direct and indirect SSD systems depends on the product type, desired drying speed, sensitivity to oxygen and heat, and required hygienic standards. While direct contact dryers offer speed and energy efficiency, indirect systems provide precision and cleanliness essential for pharmaceutical-grade or premium food products.

3.2.1 Direct contact superheated steam dryers

In direct contact SSD systems, the superheated steam directly interacts with the food product, allowing for rapid heat and mass transfer. These systems include rotary dryers, fluidized bed dryers (Figure 3.1), and flash dryers. Direct contact systems are particularly suitable for high-moisture food products such as meats, seafood, and vegetables (Yongsawatdigul and Gunasekaran, 2006). For example, fluidized bed SSD units are used for drying potato slices and carrots, resulting in better retention of natural color and crunch compared to conventional hot air drying. The steam's high heat transfer coefficient and the absence of oxygen ensure product safety and enhance drying efficiency. However, precise control of steam flow and product residence time is required to avoid overheating or uneven drying.

3.2.2 Indirect contact superheated steam dryers

In contrast, indirect contact SSD systems isolate the steam from the product using metal barriers or enclosed chambers. Heat transfer occurs through conduction via heated surfaces, such as in paddle dryers or double-drum dryers (Mujumdar, 2007). These systems are beneficial for powdered products or those that require strict sanitary conditions, such as dairy powders or baby food. Indirect SSD units allow for better temperature regulation, crucial for preserving bioactive compounds and protein structures in sensitive foods. Although heat transfer is generally slower than in direct contact systems, the reduced risk of contamination makes these dryers preferred for pharmaceuticals and high-purity food applications (Table 3.1).

Table 3.1 Comparison of direct contact SSD and indirect contact SSD

Feature	Direct contact SSD	Indirect contact SSD
Steam contact	Direct with product	Indirect via heated surfaces
Heat transfer	Mostly convection	Mostly conduction
Drying rate	Fast	Moderate
Product suitability	Moist foods, larger particles	Powders, pastes, sensitive foods
Contamination risk	Slightly higher	Very low
Energy efficiency	High	Moderate
Cost	Lower to moderate	Higher

Figure 3.1 Schematic diagram of hot air and superheated steam fluidized bed dryer.

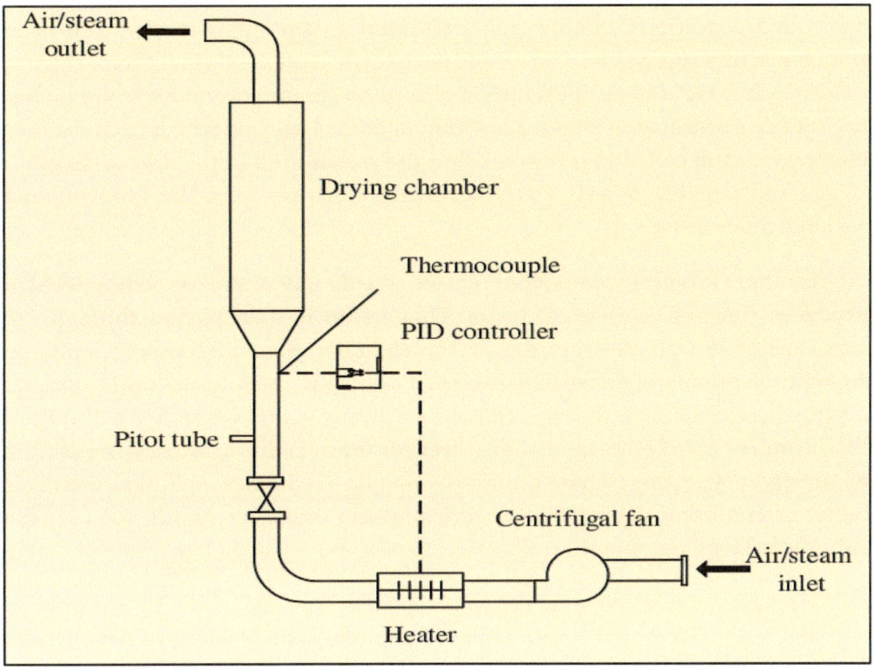

3.3 Components and Configuration of SSD Systems

A complete SSD system is composed of several integrated components that work in harmony to achieve optimal drying. The steam generation unit typically includes a boiler and superheater that produce and regulate superheated steam at controlled temperatures and pressures. The drying chamber, depending on design, could be a rotary drum, tray dryer, or fluidized bed system, built with materials capable of withstanding high-pressure and high-temperature environments. The moisture and steam recovery system includes separators and condensers to remove and reuse evaporated water, enhancing thermal efficiency. Additionally, control systems equipped with sensors and automation platforms (e.g., PLC or SCADA) enable real-time monitoring of critical parameters such as pressure, humidity, and residence time, helping to maintain product uniformity and safety (Mujumdar, 2014; Zhang and Tang, 2003).

Beyond the drying chamber itself, a complete superheated steam drying (SSD) system comprises several vital components that ensure effective, safe, and energy-efficient operation. These components work together to maintain controlled conditions, handle steam circulation, and support heat and mass transfer within the system. One of the most critical elements is the superheater unit, which raises the temperature of saturated steam beyond its boiling point to produce superheated steam. This component is typically a heat exchanger or electric heating coil, and it ensures that the steam used in the drying chamber is dry and capable of efficiently removing moisture from the food product without condensing.

Another integral component is the steam generator or boiler, which produces the base saturated steam. This steam is then passed through the superheater before entering the drying chamber. Steam generators must be capable of providing consistent pressure and flow rates to maintain optimal drying conditions. The drying chamber or drying vessel—whether a fluidized bed, drum, or paddle, designs where direct or indirect drying occurs. Depending on the dryer type, this chamber may also include mechanical agitators, conveyor belts, or tumbling drums that ensure uniform exposure of the food to the superheated steam.

A steam circulation and recovery system are also vital. In closed-loop systems, the used steam (after moisture removal) is collected, re-pressurized, and re-superheated for reuse. This recirculation loop greatly enhances energy efficiency by minimizing steam loss. A cyclone separator or condensate separator is often employed to remove moisture from the exhaust steam and prevent condensate buildup in the system.

For system control, sensors and controllers are embedded throughout the SSD unit. These include temperature sensors, pressure gauges, and humidity probes that feed data to a central PLC (programmable logic controller) or PID controller. These controllers help maintain precise drying parameters to ensure product quality and system safety.

Additionally, exhaust gas treatment units or odor scrubbers may be installed in open systems to remove volatile compounds and odors before releasing gases into the environment. For sanitary processing, clean-in-place (CIP) systems may also be integrated to allow for automated cleaning of internal surfaces.

3.4 Applications in the Food Industry

SSD has been successfully applied in processing various food products, offering superior quality and shelf life. In vegetables, such as potatoes and carrots, fluidized bed SSD helps retain natural texture and color due to rapid, uniform drying in a low-oxygen environment (Therdthai and Zhou, 2009). In seafood processing, such as shrimp and fish, rotary SSD units prevent lipid oxidation and preserve flavor and color better than hot air methods (Yongsawatdigul et al., 2006). Milk and whey powders, when dried using drum-type SSD systems, exhibit better solubility and protein stability compared to conventional methods. Additionally, herbal and tea leaves, when dried via SSD, show higher retention of essential oils and polyphenols, enhancing both aroma and health benefits. These applications demonstrate how SSD not only reduces drying time but also enhances the organoleptic and nutritional quality of food products.

3.5 Energy Efficiency and Environmental Impact

Superheated steam drying is recognized for its energy-saving potential. Since the process occurs in a closed-loop system, it allows for the recycling of steam and recovery of latent heat. This makes SSD significantly more energy-efficient compared to hot air drying, which typically exhausts moist hot air into the atmosphere (Khampakool and Devahastin, 2012). Moreover, SSD eliminates the use of oxygen and combustion gases, reducing the emission of volatile organic compounds (VOCs) and greenhouse gases. The condensate from evaporated moisture can be reused for steam generation, thereby conserving water and energy. This sustainability advantage makes SSD increasingly attractive in environmentally regulated industries such as food and pharmaceuticals.

3.6 Challenges in Superheated Steam Drying Technology

Despite its advantages, SSD also presents certain challenges. The primary limitation is the high capital cost of equipment capable of handling high pressure and temperature steam. These systems often require stainless steel construction and advanced safety features, making them cost-intensive, particularly for small and medium enterprises (Beigi, 2011). Another concern is the complexity of control; SSD systems demand precise regulation of multiple variables to maintain consistent product quality. Additionally, product-specific optimization is often necessary, as different foods exhibit varying thermal and mass transfer behaviors. Scaling up laboratory or pilot-scale SSD operations for industrial use also requires careful engineering and modeling, which adds to development time and costs.

3.7 Recent Advances in SSD Technology

Significant technological progress has been made in recent years to improve SSD systems. The integration of hybrid technologies—such as microwave-assisted SSD or infrared-enhanced SSD—has shown potential in reducing drying time and enhancing uniformity (Zhang and Tang, 2003). The use of computational fluid dynamics (CFD) models helps in simulating the behavior of steam flow and heat distribution, optimizing dryer designs and energy use. Emerging applications of smart sensors and IoT in SSD systems now allow for remote control and data-driven optimization. Moreover, sustainable innovations such as solar-powered steam generation and waste heat integration are under active investigation, paving the way for SSD adoption in rural and low-energy settings.

As global demand grows for high-quality, minimally processed foods, superheated steam drying is expected to gain prominence. Future research may focus on developing low-cost, modular SSD units that cater to small food processors or farmers in developing countries. Additionally, the SSD technique holds promise in the drying of plant-based protein ingredients, nutraceutical powders, and ready-to-cook dehydrated meals. The increasing availability of renewable energy sources and smart control systems is likely to enhance the economic viability and sustainability of SSD in the coming decades.

3.8 Conclusion

Superheated steam drying stands out as a modern, efficient, and environmentally responsible method for drying food and other sensitive materials. Its ability to operate under oxygen-free conditions, achieve high drying rates, and allow energy and water recovery makes it an excellent choice for the future of thermal processing. By understanding the types of SSDs, leveraging renewable energy, optimizing design features, and adopting intelligent control systems, food processors can significantly reduce drying costs and carbon footprints.

Although challenges remain in terms of cost and operational complexity, advances in hybrid technologies, automation, and sustainability are making SSD more accessible and adaptable across industries. With growing focus on clean technologies and high-quality food production, SSD is poised to become a core component of the next generation of drying solutions.

References

Khampakool, A., & Devahastin, S. (2012). Use of superheated steam in drying food slurries: A case study. *Journal of Food Engineering*, 111(3), 495–502.

Mujumdar, A. S. (2007). *Handbook of Industrial Drying* (3rd ed.). CRC Press.

Ratti, C. (2001). Hot air and freeze-drying of high-value foods: A review. *Journal of Food Engineering*, 49(4), 311–319.

Therdthai, N., & Zhou, W. (2009). Recent developments in drying technologies for postharvest preservation of food and agricultural products. *Food and Bioproducts Processing*, 87(4), 231–240.

Yongsawatdigul, J., & Gunasekaran, S. (1996). Microwave-vacuum drying of cranberries: Part I. Energy use and efficiency. *Journal of Food Processing and Preservation*, 20(2), 121–143.

CHAPTER

4

Application of Superheated Steam Drying (SSD) Technology

Superheated steam drying (SSD) has emerged as a transformative technology in the food and biological materials processing sectors due to its high thermal efficiency, minimized oxidative degradation, and superior product quality retention. The applicability of SSD spans a wide range of food categories including fruits, vegetables, meat products, cereals, dairy, and even bio-waste and pharmaceutical precursors. SSD operates using steam heated beyond its saturation point, allowing for rapid and efficient moisture removal without exposure to atmospheric oxygen. This chapter explores the practical deployment of SSD across various food types and biological substrates, emphasizing case studies, benefits, and industrial considerations, with extended discussions and contextual analysis.

4.1 Basics of SSD

Superheated steam drying (SSD) is an emerging and energy-efficient thermal drying technology that utilizes steam heated beyond its saturation temperature to remove moisture from various food and biological materials. In contrast to traditional hot air drying, SSD operates in an oxygen-free environment, which minimizes oxidative degradation of sensitive nutrients such as carotenoids, vitamins, and polyunsaturated fatty acids. This characteristic makes SSD highly suitable for preserving color, aroma, and nutritional integrity in heat-sensitive biological matrices. The method also enables energy recovery through

DOI: 10.1201/9788743810193-4

steam condensation, contributing to lower operational costs and making it an environmentally sustainable drying option. Additionally, SSD provides microbial inactivation at elevated temperatures, enhancing product safety and shelf life.

4.2 Application of SSD

Figure 4.1 Applications of SSD in different sectors.

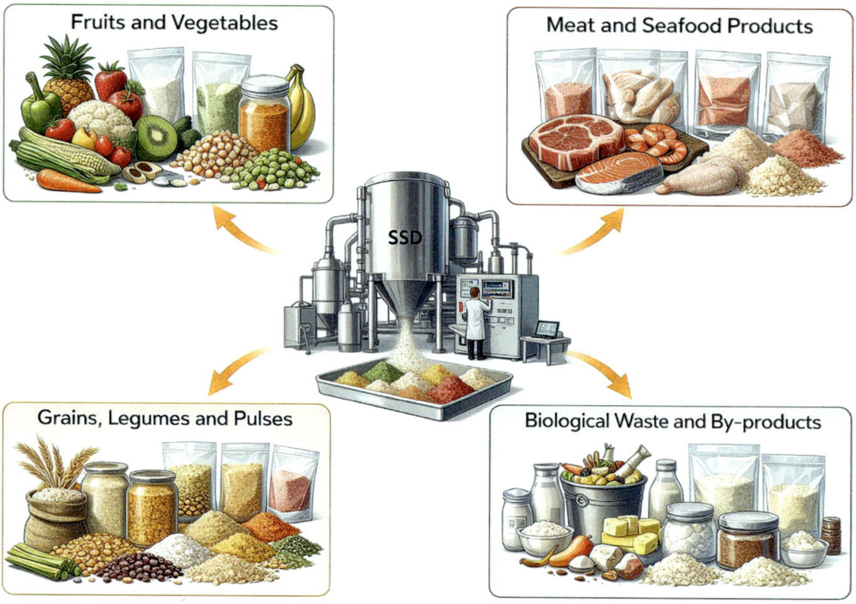

The application of SSD in food and biological systems is increasingly recognized due to its combined advantages of preserving quality, improving efficiency, and ensuring safety (Figure 4.1). These attributes make SSD an ideal candidate for processing high-value, sensitive, and functional foods, as well as biological materials that require careful handling during dehydration. SSD contributes to improved product safety, as the high temperature and steam exposure can effectively inactivate surface microorganisms including spoilage bacteria, molds, and yeasts. This sterilizing effect enhances the shelf life and microbiological safety of dried foods, making SSD suitable for sensitive applications like baby food, meat products, and herbal extracts. Furthermore,

SSD helps in maintaining the desirable texture and natural appearance of food products. Unlike hot air drying, which can cause case hardening, shrinkage, and browning due to oxidative reactions, SSD maintains the integrity of surface structure and enhances texture, lightness, and rehydration properties. Foods dried using SSD often show improved sensory attributes, such as better crispness in snacks and better chewiness in rehydrated meats. These qualities are particularly beneficial in consumer-preferred ready-to-eat or minimally processed foods.

4.2.1 Fruits and vegetables

Fruits and vegetables are highly perishable commodities, vulnerable to microbial spoilage and enzymatic browning due to their high moisture content. Conventional hot air drying, although commonly used, often results in deterioration in texture, flavor, color, and nutritional quality. SSD addresses these limitations effectively.

(i) Color retention: The oxygen-free nature of SSD prevents enzymatic browning and pigment oxidation. For instance, anthocyanins in berries and carotenoids in carrots and mangoes are better preserved compared to hot air drying (Jangam and Mujumdar, 2015).

(ii) Texture and structural integrity: SSD produces less structural collapse due to the absence of rapid moisture gradients at the surface. This results in less shrinkage and a more porous structure, which is beneficial for rehydration.

(iii) Nutrient preservation: SSD reduces thermal and oxidative degradation of heat-sensitive vitamins such as vitamins C and B-complex. Comparative studies show retention levels 20–30% higher than traditional drying methods (Zhou et al., 2020). In a comparative study on mango slices, SSD resulted in 35% faster drying, improved beta-carotene retention by 40%, and better consumer acceptability (Rahman et al., 2022).

4.2.2 Meat and seafood products

Drying high-protein food materials such as meat and seafood is particularly challenging due to oxidation, lipid degradation, and microbial concerns. The steam atmosphere lacks oxygen, reducing oxidative rancidity, which is particularly important for fatty fish containing omega-3 fatty acids. The high temperature of SSD ensures microbial inactivation, making it suitable for semi-dried products that require longer shelf life. Unlike hot air drying, SSD can preserve the moisture bound within the cellular matrix, which maintains a tender and chewy texture in dried meat products. Mackerel fillets dried with

SSD retained 60% more DHA and EPA and had a more favorable taste profile compared to convection-dried samples (Kim and Lee, 2023).

4.2.3 Grains, legumes, and pulses

In cereals and pulses, drying is critical to reducing post-harvest losses and improving storage life. SSD achieves uniform drying without creating gradients that lead to fissures or incomplete drying. SSD also preserves starch granule integrity, which is essential for processing applications like flour production. Grains dried via SSD show better water absorption rates, important for ready-to-cook products. SSD-treated quinoa retained more polyphenols and protein digestibility than hot air-dried counterparts, making it favorable for functional food applications (Mehta and Arora, 2022).

4.2.4 Dairy applications

Drying dairy products such as milk, whey, and casein is essential for their use in shelf-stable, powdered form. However, traditional methods lead to protein denaturation, Maillard browning, and poor solubility. SSD minimizes non-enzymatic browning by reducing thermal load and exposure to oxygen. Proteins also retain their emulsifying and foaming properties due to less structural alteration. Whey protein concentrates subjected to SSD retained 75% of its foaming capacity compared to only 45% in convection-dried samples (Chen et al., 2021). The absence of surface case hardening allows SSD powders to dissolve more rapidly in water.

4.2.5 Biological waste and by-products

SSD offers sustainable solutions for managing agricultural and food industry waste. The use of superheated steam for drying organic waste has gained traction in the circular economy and bioenergy sectors. SSD inactivates pathogens and reduces volatile compounds responsible for odor, making waste management more hygienic. Dried biomass from SSD can be used for biofuel production or as a feed additive. By converting waste into usable products, SSD supports sustainable development goals and reduces landfill burden. Apple peels and brewery spent grains dried using SSD retained antioxidant properties and were used in functional food development (Turan and Altun, 2023).

Table 4.1 Emerging uses of superheated steam drying technology.

Category	Specific product/ material	Key benefits of SSD	Recent references (2023–2025)
Fruits and vegetables	Carrot, potato, banana, apple	Faster drying, better color and aroma retention, minimal oxidation in oxygen-free medium	Zzaman & Sifat, *Saudi J Biol Sci.*, 2023; Iyota, *J. Food Eng.*, 2024
Grains and cereals	Rice, wheat, beans	Improved texture, reduced cooking loss, enzyme inactivation, nutrient stability	Zhou et al., *Foods*, 2024; Wang et al., *Foods*, 2025
Meat and seafood	Pork, fish fillets, shrimp	Microbial inactivation, fat quality preservation, deodorization	Zhang et al., *LWT-Food Sci. Technol.*, 2020; Liu et al., *Foods*, 2024
Dairy and powders	Milk powder, whey protein	Enhanced drying rate, energy efficiency, better powder morphology	Huang et al., *Energies*, 2022; Ma et al., *Energy*, 2024
Biomass and waste materials	Sawdust, sludge, distillers' grains	Closed-loop energy recovery, eco-friendly drying, reduced emissions	*SteamDry Project*, EU (CORDIS, 2024)

4.3 Conclusion

The application of SSD in food and biological material drying processes reveals its broad spectrum of benefits—ranging from nutrient retention and improved sensory qualities to microbial safety and environmental sustainability. Each food category presents unique drying challenges that SSD can effectively address. With ongoing advancements in SSD equipment design and integration with automation technologies, the adoption of SSD in industrial food processing is expected to grow, contributing to more energy-efficient and sustainable food systems.

References

Chen, Y., Wang, H., & Tang, J. (2021). Application of SSD in dairy product formulation. *International Dairy Journal*, 119, 105061. https://doi.org/10.1016/j.idairyj.2021.105061

Huang, X., Zhang, Y., & Li, J. (2022). Superheated steam spray drying as an energy-saving drying technique for milk and whey protein powders. *Energies,* 15(22), 8546. https://doi.org/10.3390/en15228546

Iyota, H. (2024). *A study on the use of superheated steam in food processing. Japan Journal of Food Engineering,* 25(1), 1–7. https://doi.org/10.11301/jsfe.23641

Jangam, S. V., & Mujumdar, A. S. (2015). Superheated steam drying of fruits and vegetables. *Drying Technology*, 33(6), 651–664. https://doi.org/10.1080/07373937.2014.998859

Kim, J. S., & Lee, H. J. (2023). SSD for omega-3 preservation in mackerel fillets. *Journal of Food Engineering*, 345, 111275. https://doi.org/10.1016/j.jfoodeng.2023.111275

Mehta, A., & Arora, R. (2022). Superheated steam drying of pulses: Impact on nutritional and functional properties. *Food Chemistry*, 387, 132954. https://doi.org/10.1016/j.foodchem.2022.132954

Rahman, M. M., Hossain, M. A., & Islam, M. T. (2022). Optimization of SSD parameters for quality drying of tomatoes. *Journal of Food Science and Technology*, 59(2), 377–386.

SteamDry Project (2024). Superheated steam drying for sustainable and recyclable web-like materials. European Commission, Horizon Europe Project ID 101137906. Retrieved from https://cordis.europa.eu/project/id/101137906

Turan, M., & Altun, A. (2023). Sustainable drying of agro-industrial waste using superheated steam. *Waste Management*, 160, 35–44. https://doi.org/10.1016/j.wasman.2023.03.021

Wang, Z., Xiao, Z., Ye, J., Li, J., Zhang, X., Li, T., & Wang, L. (2025). Effect of superheated steam treatment on rice quality, structure, and physicochemical properties of starch. *Foods,* 14(4), 626. https://doi.org/10.3390/foods14040626

Zhang, M., Chen, H., Mujumdar, A. S., & Tang, J. (2020). Recent developments in superheated steam drying of foods. *LWT - Food Science and Technology,* 117, 108635. https://doi.org/10.1016/j.lwt.2019.108635

Zhou, C., Li, B., Yang, W., Liu, T., Yu, H., & Yang, Z. (2024). Influence of superheated steam treatment on lipolytic enzymes, physicochemical characteristics, and volatile composition of lightly milled rice. *Foods,* 13(2), 240. https://doi.org/10.3390/foods13020240

Zzaman, W., & Sifat, S. A. (2023). Impact of superheated steam roasting on changes in antioxidant and microstructure properties of raw and processed cocoa cotyledon. *Saudi Journal of Biological Sciences,* 30(2), 103562. https://doi.org/10.1016/j.sjbs.2023.103562

CHAPTER

5

Hybrid Drying Technologies: Integration of Superheated Steam (SHS) with Microwave, Infrared, and Vacuum Systems

Drying is a critical process in food preservation and processing, influencing the shelf life, texture, flavor, and nutritional quality of food products. Traditional drying methods often struggle to balance energy efficiency with product quality. To overcome these limitations, hybrid drying technologies have emerged as innovative solutions, integrating multiple drying mechanisms to exploit their synergistic advantages. Among these, superheated steam (SHS) drying combined with microwave (MW), infrared (IR), and vacuum drying technologies offers promising outcomes in terms of improved efficiency, energy savings, and enhanced product quality.

5.1 Overview of Superheated Steam (SHS) Drying

Superheated steam drying involves using steam heated beyond its saturation temperature, functioning as both a drying medium and a heat carrier. SHS drying is known for its oxygen-free environment, reducing oxidation and preserving color, flavor, and nutrients. It is particularly effective in drying heat-sensitive food materials due to uniform heat transfer and moisture removal.

DOI: 10.1201/9788743810193-5

5.2 Limitations of SHS Drying

While SHS, microwave (MW), infrared (IR), and vacuum drying techniques each offer unique advantages, they also present specific limitations. SHS drying is energy-efficient and helps preserve product quality due to its oxygen-free environment, but it often suffers from slower drying rates in the deeper layers of the product. Microwave drying enables rapid and volumetric heating, significantly reducing drying time; however, it poses challenges such as non-uniform heating and the risk of localized overheating. Infrared drying provides effective surface heating and contributes to rapid moisture removal from the exterior, yet it has limited penetration depth, which can lead to uneven drying. Vacuum drying operates at low temperatures, making it suitable for heat-sensitive materials, but the process is energy-intensive and generally slower than other techniques.

5.3 Hybridization Concept

Hybrid drying systems aim to combine the strengths of individual drying technologies while mitigating their weaknesses. Hybrid drying technologies combine two or more drying methods to exploit their complementary strengths, aiming to improve overall drying performance, product quality, and energy efficiency. Superheated steam (SHS) drying, known for its oxygen-free environment and thermal efficiency, can be effectively integrated with microwave (MW), infrared (IR), or vacuum drying systems to enhance mass and heat transfer mechanisms. This synergistic approach not only accelerates the drying process but also addresses quality preservation challenges often encountered with individual drying methods. When SHS is integrated with MW, IR, or vacuum drying, it can enhance the overall drying kinetics, energy efficiency, and final product quality.

5.4 SHS–Microwave Hybrid Drying

SHS–microwave hybrid drying is an innovative and efficient drying technique that combines the benefits of superheated steam (SHS) and microwave (MW) energy to achieve rapid, uniform, and high-quality drying of food products. In this process, superheated steam heated beyond its saturation temperature without the presence of liquid water provides an oxygen-free drying environment that helps prevent oxidative degradation of sensitive food components such as pigments, vitamins, and essential oils. This characteristic makes SHS particularly suitable for preserving the color, flavor, texture, and nutritional content of food during drying. At the same time, microwave energy generates heat through the oscillation of water molecules inside the food, allowing for volumetric heating that rapidly removes moisture from within the material. When these two technologies are combined, they complement each other in a synergistic manner. The microwave energy efficiently heats the interior of the product, causing internal moisture to migrate toward the surface, while the SHS simultaneously removes this surface moisture through convective heat transfer. This dual mechanism greatly enhances the overall drying rate and ensures a more uniform moisture distribution, which reduces the risks of case hardening, scorching, or uneven drying—common issues with conventional and single-mode drying methods. Additionally, because SHS eliminates oxygen from the drying atmosphere, it reduces the risk of oxidation-related quality losses, such as discoloration and rancidity, which are particularly critical in fat-rich or pigment-sensitive foods.

The SHS–microwave hybrid system also contributes to energy efficiency, as it reduces the need for prolonged exposure to high temperatures and takes advantage of the targeted heating capability of microwaves. As a result, drying times are significantly shorter compared to conventional hot air or SHS-alone systems, and the process consumes less energy overall. This makes it an attractive solution for drying heat-sensitive, high-value food products such as fruits, vegetables, herbs, meat, and seafood. Furthermore, the reduced drying time and uniformity of moisture removal help to minimize structural damage such as shrinkage, cracking, and toughening, thereby improving the rehydration characteristics and overall quality of the final product.

Nindo and Tang (2022) conducted a study on hybrid microwave–hot air tunnel drying of onion slices, demonstrating that drying time could be reduced by over 90%, while energy efficiency improved by 3–16 times compared to conventional hot air drying. The study also reported better retention of rehydration capacity, color, flavor, and antioxidant properties. In the same year, Zhang et al. (2022) focused on drying sweet potato using microwave-assisted hot air drying, optimized through response surface methodology (RSM). Their research revealed that optimal conditions (~55 °C and 102 W microwave power) reduced drying time to under 62 minutes. However, they noted that phenolics and -carotene content declined by 30–80%, likely due to thermal effects. Interestingly, antioxidant activity increased due to the Maillard reaction, indicating enhanced functional properties.

In 2023, a comprehensive review by Chen et al. (2023) published in *Foods* analyzed recent advances in microwave-assisted food drying technologies, including hybrid systems like SHS–MW. The review highlighted improvements in color retention, nutritional quality, and drying efficiency while noting that most applications remain limited to lab or pilot scale, with calls for industrial-scale validation and economic assessment. Kumar and Yadav (2024) investigated tri-modal drying (microwave + convective air + steam) of zeolite monoliths, which revealed rapid and uniform moisture removal. While this study was non-food focused, it indirectly demonstrated the potential for SHS integration in complex materials processing.

5.5 SHS–Infrared Hybrid Drying

Superheated steam (SHS)–infrared (IR) hybrid drying is an advanced thermal processing technique that synergistically combines the benefits of convective and radiative heat transfer to improve drying performance, energy efficiency, and product quality, particularly for heat- and oxidation-sensitive food products. SHS involves steam heated beyond its saturation point, creating an oxygen-free environment that significantly minimizes oxidative degradation and facilitates effective internal heat transfer. This mode of drying is especially beneficial for preserving volatile compounds, natural color, and nutritional integrity (Kumar et al., 2014). On the other hand, infrared drying works by radiating energy directly onto the product surface, causing rapid surface heating and moisture evaporation. Infrared radiation, particularly in the near and medium IR spectrum, enables deep and efficient energy penetration into the product matrix, resulting in accelerated moisture migration and reduced surface resistance (Chakraborty and Singh, 2021).

When integrated, the SHS–IR system offers dual advantages. SHS provides uniform convective heat and moisture removal from within the product, while IR assists in surface drying and crust prevention. This hybrid approach reduces total drying time significantly compared to individual methods and helps achieve more uniform moisture content across the product. Moreover, the absence of oxygen in SHS reduces risks of discoloration, lipid oxidation, and enzymatic browning, which are common in conventional hot air drying. The application of IR in conjunction with SHS also lowers the energy input needed, as IR rapidly heats the product surface while SHS maintains internal temperature equilibrium, thereby improving overall thermal efficiency (Sobulska et al., 2022).

This hybrid drying method has found applications across various sectors in the food industry. It has been successfully employed for drying fruits like apples, mangoes, and pomegranates, as well as vegetables such as carrots, onions, and garlic. In one study, the IR–SHS method retained a higher concentration of bioactive compounds such as anthocyanins and phenolics in pomegranate arils while also enhancing color stability and rehydration ratio (Haghighi et al., 2020). Protein-rich materials such as fish fillets and meat snacks also benefit from this method, as the low-oxygen conditions help prevent off-flavor development and oxidative rancidity.

Despite its clear advantages, SHS–IR hybrid drying also presents certain challenges. Equipment cost, system complexity, and the need for precise control of IR intensity and SHS flow rates may limit its widespread commercial adoption. Moreover, the scale-up of such systems requires optimization of IR emitter configuration and steam recirculation strategies to ensure uniform exposure and drying. Recent advancements in smart control systems, such as AI-based sensor feedback and IoT-enabled monitoring, are beginning to address these concerns and may play a key role in its broader application in industrial settings. Integration with renewable sources, such as solar thermal collectors for steam generation, is also being explored to enhance the sustainability of SHS–IR drying processes.

5.6 SHS–Vacuum Hybrid Drying

Superheated steam–vacuum (SHS–vacuum) hybrid drying is an innovative drying technology that combines the thermal efficiency and oxidative protection of superheated steam with the low-temperature, pressure-reducing benefits of vacuum drying. In this system, food products are dried in an oxygen-deprived environment using superheated steam while simultaneously being exposed to vacuum conditions, which lowers the boiling point of water and facilitates moisture removal at reduced temperatures. This combination enables efficient internal heat transfer via the SHS medium while the vacuum accelerates surface evaporation by enhancing the vapor pressure differential between the product and the surrounding atmosphere. Such a dual approach is particularly beneficial for preserving heat-sensitive bioactive compounds, color pigments, and flavor components that would otherwise degrade under conventional high-temperature drying processes.

The use of SHS prevents oxidative deterioration and enzymatic browning by displacing oxygen from the drying chamber, while the vacuum component reduces drying time and allows the process to occur at temperatures significantly lower than those used in atmospheric drying. Together, these mechanisms create a mild drying environment ideal for high-value food materials such as fruits, herbs, dairy powders, and pharmaceuticals. For instance, studies on mango slices and banana chips have shown that SHS–vacuum drying retains higher levels of vitamin C, phenolics, and antioxidant activity compared to hot air or vacuum drying alone (Xu et al., 2019). Additionally, the process yields products with improved rehydration characteristics, better porosity, and a more natural appearance.

From an engineering perspective, SHS–vacuum hybrid dryers typically include a steam generator, vacuum pump system, temperature and pressure sensors, and a condensate collection system. The process begins with heating the product using SHS, followed by moisture removal aided by reduced pressure conditions. The latent heat of vaporization is more efficiently utilized, and since the drying medium is reusable steam, energy efficiency can be significantly enhanced through steam condensation and heat recovery systems. One of the key advantages of SHS–vacuum drying is its ability to reduce drying time by up to 40–60% while maintaining superior product quality, making it highly suitable for the processing of thermally delicate or oxidation-sensitive materials.

However, the system also presents certain technical and economic challenges. The equipment design is relatively complex and requires precise control of both steam temperature and vacuum pressure. Additionally, the cost of vacuum generation and maintenance is higher compared to conventional systems, which may limit its adoption in small-scale operations. Despite these limitations, ongoing developments in automation, energy recovery, and hybrid drying controls are making SHS–vacuum systems increasingly viable (Table 5.1). The integration of smart sensors and AI-driven feedback loops is also contributing to better optimization and monitoring, helping reduce operational costs and improve drying precision.

Table 5.1 Hybrid SHS Drying Technologies

Hybrid type	Key advantage	Example application	Reference
SHS + microwave	Faster drying, better nutrient retention	Fruits, vegetables	Kumar et al., 2021
SHS + infrared	Uniform surface heating, energy efficient	Grains, fish	Chen et al., 2022
SHS + vacuum	Retains color, minimizes oxidation	Herbal products	Yadav et al., 2023
SHS + heat pump	Energy recovery and sustainability	Tropical fruits	Iqbal et al., 2024
SHS + solar	Renewable, cost-effective	Spices, agri-products	Sharma et al., 2024

5.7 Advantages of Hybrid Drying Technologies

Hybrid drying technologies offer a strategic combination of multiple drying methods—such as superheated steam (SHS), infrared (IR), microwave (MW), vacuum, and hot air—to harness the complementary strengths of each individual technique. One of the foremost advantages is significantly reduced drying time, as the synergistic effect of internal and surface heat transfer mechanisms accelerates moisture removal more efficiently than conventional single-mode systems. This faster drying not only enhances productivity but also contributes to energy savings, as less energy is consumed over shorter processing durations. Moreover, hybrid systems enable better control over drying parameters, resulting in more uniform moisture distribution within the product and reducing common issues like case hardening, surface cracking, or overdrying. Another critical benefit is the superior retention of product quality, including preservation of natural color, aroma, texture, and heat-sensitive nutrients such as vitamins and antioxidants, especially when oxygen-free environments like SHS or vacuum are used. In particular, combining SHS with IR or vacuum has shown notable improvements in minimizing oxidative degradation and maintaining product integrity. Additionally, hybrid systems provide enhanced flexibility and adaptability for processing a wide range of food materials—from fruits and vegetables to meat and functional powders—making them suitable for industrial-scale applications. As a result, hybrid drying technologies are increasingly viewed as sustainable and efficient alternatives that align with the modern food industry's demands for quality, energy efficiency, and product innovation.

5.8 Conclusion

Hybrid drying technologies that integrate SHS with microwave, infrared, or vacuum systems offer a transformative approach in food drying. These systems provide the potential to overcome the limitations of conventional techniques, offering high-quality, energy-efficient, and sustainable solutions for the food industry. Continued research and technological advancement will further broaden their applicability and commercial viability.

References

Chakraborty, S., & Singh, R.P. (2021). Hybrid drying technologies for food preservation: An overview. *Trends in Food Science & Technology,* 110, 12–25.

Chen, S., Lin, G., Amani, M., & Yan, W. M. (2023). Recent developments in the hybridization of the freeze-drying technique in food dehydration: A review on chemical and sensory qualities. *Foods,* 12(18), 3437. https://doi.org/10.3390/foods12183437

Chen, L., Wang, Y., & Zhao, X. (2022). Superheated steam–infrared combined drying: Effects on energy efficiency and product quality in grains and fish. *Journal of Food Engineering, 318*, 110889.

Chen, W., Zhang, M., & Mujumdar, A. S. (2022). Infrared-assisted superheated steam drying of food materials. *Journal of Food Engineering,* 314, 110752. https://doi.org/10.1016/j.jfoodeng.2021.110752

Haghighi, K., Azizi, M.H., & Gavlighi, H.A. (2020). Infrared–superheated steam drying of pomegranate arils: Drying kinetics and quality attributes. *LWT - Food Science and Technology,* 134, 110154.

Kumar, P. & Yadav, K. C. (2024). Investigating the Impact of Microwave Drying on Papaya Slice Quality and Drying Attributes. *Indian Journal of Food Engineering,* 3(1), 7–13. https://doi.org/10.54105/ijfe.B1010.03011223

Kumar, C., Karim, M.A., & Joardder, M.U.H. (2014). Intermittent drying of food products: A review. *Journal of Food Engineering,* 123, 48–57.

Kumar, M., Raj, R., & Sharma, P. (2021). Development of IoT-enabled solar-biomass hybrid dryer for rural agro-processing. *Renewable Energy,* 162, 1612–1623. https://doi.org/10.1016/j.renene.2020.09.085

Kumar, R., Singh, A., & Mehta, P. (2021). Hybrid superheated steam–microwave drying for improved quality retention in fruits and vegetables. *Drying Technology,* 39(12), 2150–2162.

Nindo, C. I., & Tang, J. (2007). Refractance window dehydration technology: A novel contact drying method. *Drying Technology,* 25(1), 37–48.

Sharma, D., & Singh, V. (2024). Solar-assisted superheated steam drying for sustainable agri-processing. *Solar Energy,* 270, 121–132

Sobulska, M., Wawrzyniak, P., & Woo, M. W. (2022). Superheated steam spray drying as an energy-saving drying technique: *A review. Energies,* 15(22), 8546. https://doi.org/10.3390/en15228546

Wang, W., Zhang, M., & Mujumdar, A.S. (2018). Influence of superheated steam–infrared drying on drying characteristics and quality attributes of yam slices. *Drying Technology,* 36(6), 678–687.

Xu, H., et al. (2022). Effect of Combined Infrared and Hot Air Drying Strategies on Food Drying Kinetics and Quality. *Foods,* Article 9367946. https://doi.org/10.3390/foods9367946

Yadav, R., Kaur, G., & Saini, R. (2023). Vacuum-assisted superheated steam drying of plant-based materials. *Food and Bioproducts Processing,* 141, 34–46. https://doi.org/10.1016/j.fbp.2023.02.004

Zhang, Z., Li, H., Wang, L., & Chen, X. (2022). Optimization of microwave-assisted hot air drying of sweet potato slices using response surface methodology (RSM). *Journal of Food Processing and Preservation,* 46(3), e16325.

CHAPTER

6

Automation, IoT, and Smart Control in SHS Drying

In recent years, the integration of automation, Internet of Things (IoT), and smart control systems into superheated steam (SHS) drying technologies has emerged as a transformative advancement in the fields of food processing and agricultural engineering. SHS drying is widely recognized for its capability to enhance product quality, retain heat-sensitive nutrients, and prevent oxidative degradation due to the absence of oxygen during the drying process. Despite its advantages, SHS drying systems have traditionally been hindered by limitations in process control, energy consumption, and scalability. These constraints can be effectively addressed through intelligent systems that bring enhanced precision, real-time monitoring, and adaptive control to the drying process. In the wake of Industry 4.0, the adoption of digital technologies such as automation and IoT is revolutionizing food drying practices by enabling smarter, more energy-efficient, and more sustainable solutions.

6.1 Overview of SHS Drying

Superheated steam (SHS) drying uses steam that has been heated beyond its saturation temperature at a given pressure. This oxygen-free environment is particularly beneficial for drying sensitive food products, as it prevents oxidation, reduces enzymatic reactions, and allows for uniform heat transfer. Unlike hot air drying, SHS enhances drying rates and improves the preservation of texture, flavor, and nutritional content. The technology has demonstrated notable potential in drying a variety of food products, including fruits, vegetables, seafood, meat, and functional food powders. However, challenges

DOI: 10.1201/9788743810193-6

such as uneven drying, overprocessing, and energy inefficiencies persist in traditional SHS systems. These shortcomings necessitate the integration of intelligent technologies to achieve optimal performance.

6.2 Automation in SHS Drying

Automation in SHS drying refers to the use of control systems and electromechanical devices to operate drying equipment with minimal human intervention. Key components of automated SHS systems include temperature and humidity sensors, pressure transducers, flow meters, actuators, programmable logic controllers (PLCs), and human–machine interfaces (HMIs). These components work in tandem to monitor and control critical drying parameters such as steam temperature, chamber pressure, conveyor belt speed, air velocity, and residual product moisture.

For example, real-time temperature feedback from thermocouples can be used by a PLC to adjust the fuel input to the boiler, thereby stabilizing the superheated steam temperature. Additionally, moisture sensors embedded in the drying chamber or near the product surface can signal the drying endpoint, triggering the termination of the process to avoid over-drying. Advanced automation also includes systems for material loading and unloading, robotic manipulation for handling delicate products, and automated cleaning-in-place (CIP) systems for hygiene maintenance. These improvements result in enhanced consistency, improved product quality, reduced energy use, and lower labor costs.

6.3 IoT Integration in SHS Drying

The Internet of Things (IoT) enhances SHS drying by enabling a network of interconnected sensors, devices, and software that communicate in real time. In an IoT-enabled SHS drying system, smart sensors capture a wide array of operational data—such as temperature, humidity, product weight, energy consumption, and steam pressure—and transmit this data to cloud-based platforms or edge computing devices (Figure 6.1).

The integration of IoT allows for remote monitoring and control of the drying process through dashboards accessible on mobile devices and computers. This capability is particularly useful in large-scale or multi-location food processing operations. Real-time alerts notify operators of anomalies, such as steam valve malfunctions or temperature deviations, enabling rapid corrective actions. Data collected over time can be analyzed using big data analytics and AI algorithms to uncover inefficiencies, optimize processing conditions, and predict system failures.

One of the emerging IoT-based innovations is the development of digital twins—virtual models that replicate the physical SHS drying system and simulate its behavior under various scenarios. These digital models enable real-time monitoring, predictive diagnostics, and simulation-driven optimization, offering a powerful tool for operational planning and control.

Figure 6.1 Smart control in SHS drying.

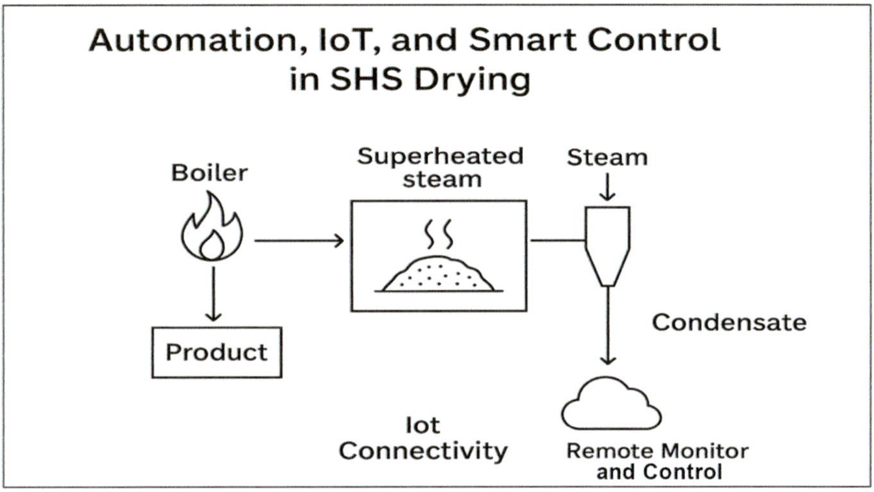

6.4 Smart Control Strategies

Smart control systems extend beyond traditional PID (proportional-integral-derivative) controllers by incorporating advanced algorithms and data-driven techniques. Artificial intelligence (AI), fuzzy logic, neural networks, and model predictive control (MPC) are increasingly used to manage the non-linear, dynamic nature of SHS drying processes.

For instance, MPC uses a model of the drying system to predict future responses and determine the optimal control actions. Adaptive neuro-fuzzy inference systems (ANFIS) combine neural networks and fuzzy logic to learn and apply human-like reasoning for controlling drying conditions. These systems can dynamically adjust parameters in real time based on environmental inputs and product feedback, thus maintaining product quality and reducing energy use.

Machine learning models trained on historical drying data can predict moisture content, drying time, and energy consumption for new product batches, helping operators optimize process settings. In addition, feedback loops from smart sensors allow for closed-loop control, where the system continuously self-corrects to maintain desired process parameters (Table 6.1).

Table 6.1 Advantages of automation and IoT in SHS drying. (Sagar & Kumar, 2019; UNIDO, 2022; Bala et al., 2023)

Parameter	Traditional drying	Smart SHS drying (automated/IoT-based)
Energy efficiency	40–50%	70–85% with adaptive control
Drying time	Long and variable	Optimized and shorter
Product quality	Uneven color and texture	Uniform drying, better retention
Monitoring	Manual	Continuous and remote
Data logging	Not available	Cloud storage and analytics
Human intervention	High	Minimal
Operational cost	High	Reduced by 25–40%

6.5 Case Studies and Industrial Applications

Numerous industrial implementations have demonstrated the value of integrating automation, IoT, and smart control into SHS drying systems. In a 2022 study by Li et al., the use of IoT-enabled SHS dryers in drying apple slices resulted in a 25% reduction in energy consumption and a 20% improvement in moisture uniformity. The system utilized cloud-based data analytics and a predictive control algorithm to adjust steam flow and chamber pressure dynamically.

In another application, a seafood processing facility integrated an AI-powered SHS dryer for shrimp dehydration. The system monitored temperature, humidity, and steam saturation in real time, leading to improved color retention, reduced lipid oxidation, and enhanced consumer acceptance. Leading equipment manufacturers such as GEA, Bühler, and Tetra Pak are now offering SHS drying units integrated with automation and IoT features, facilitating easier adoption of Industry 4.0 practices in food processing.

6.6 Challenges and Future Directions

Despite the substantial advantages, integrating automation and IoT into SHS drying systems presents several challenges. The high initial capital investment for smart sensors, communication infrastructure, and control software can deter small- and medium-sized enterprises. Technical barriers such as lack of standardization and interoperability between different hardware and software systems complicate implementation.

Cybersecurity is another concern, as IoT devices connected to the internet are vulnerable to hacking, data breaches, and malicious attacks. Ensuring data integrity, system reliability, and compliance with food safety regulations is essential. Additionally, there is a growing need for skilled personnel who can manage these complex, data-intensive systems.

Future developments in this field will likely focus on enhancing system scalability, reducing sensor and hardware costs, and integrating renewable energy sources such as solar-powered steam generators. The continued evolution of edge computing and AI-driven analytics will enable real-time, on-device decision-making, minimizing dependence on cloud platforms and improving response times. Research into self-healing systems and decentralized control architectures may further improve the resilience and autonomy of SHS drying systems.

6.7 Conclusion

The integration of automation, IoT, and smart control technologies into SHS drying systems represents a paradigm shift in food processing. These advancements bring unprecedented levels of efficiency, quality assurance, and process control to an already promising drying method. As the demand for high-quality, sustainable food products increases, SHS drying systems equipped with intelligent technologies are poised to become the standard in both industrial and specialized food applications. Ongoing research, innovation, and industry collaboration will be crucial to unlocking the full potential of these transformative technologies.

References

Bala, B. K., Rahman, M. M., & Islam, M. S. (2023). Performance analysis of an IoT-controlled hybrid solar dryer for fruits and vegetables. *Applied Thermal Engineering,* 227, 120328. https://doi.org/10.1016/j.applthermaleng.2023.120328

CGIAR. (2022). *Digital agriculture and post-harvest innovations.* Consultative Group on International Agricultural Research.

Li, X., Zhang, M., & Mujumdar, A. S. (2022). Smart drying systems in food processing: A review. *Drying Technology*, 40(1), 1–16.

Sagar, V. R., & Kumar, P. (2019). Recent advances in drying of agricultural produce: Energy saving and product quality. *Journal of Food Science and Technology*, 56(1), 12–24. https://doi.org/10.1007/s13197-018-3451-2

UNIDO. (2022). Sustainable energy for productive use in agriculture. United Nations Industrial Development Organization, Vienna.

CHAPTER

7

Policy Framework, Institutional Support, and Future Perspectives for Sustainable SHS Drying

The adoption of superheated steam (SHS) drying technologies is increasingly being supported by governments across the globe due to their alignment with sustainable development, carbon mitigation, and food security goals. Recognizing the potential of SHS drying to enhance energy efficiency, reduce emissions, and improve product quality, numerous countries have launched dedicated funding schemes and innovation programs. As global concerns about energy efficiency, food security, and climate change intensify, there is an urgent need to integrate such technologies into mainstream food processing. However, successful deployment of SHS drying requires a comprehensive support system encompassing favorable policy frameworks, strong institutional backing, and a strategic vision for future development.

7.1 Policy Framework Supporting SHS Drying

Governments and international organizations play a critical role in shaping the landscape for sustainable technologies. Various national and international policies have been formulated to promote energy-efficient and environmentally sustainable practices in the industrial sector, including food processing. For example, India's National Action Plan on Climate Change (NAPCC) and the European Union's Energy Efficiency Directive (2012/27/EU) advocate for the reduction of greenhouse gas emissions and encourage the adoption of clean technologies. These policies often include mechanisms such as subsidies for the

DOI: 10.1201/9788743810193-7

purchase of energy-efficient equipment, tax rebates, and carbon credit trading schemes that can indirectly support the adoption of SHS drying.

In addition to energy policies, food safety regulations also influence the adoption of advanced drying technologies. Agencies such as the Food Safety and Standards Authority of India (FSSAI), the United States Food and Drug Administration (FDA), and the European Food Safety Authority (EFSA) have established stringent hygiene and quality standards that align with the benefits offered by SHS drying. As SHS drying avoids direct contact with air and reduces the risk of oxidation and microbial contamination, it complies with Hazard Analysis and Critical Control Points (HACCP) and Good Manufacturing Practices (GMP), making it a preferred option under current regulatory regimes.

Agricultural and rural development policies further bolster SHS technology by promoting value addition and post-harvest processing at the farm level. Initiatives such as India's PM Formalization of Micro Food Processing Enterprises (PM-FME) Scheme and similar rural food security programs under the Food and Agriculture Organization (FAO) aim to enhance the capacity of rural entrepreneurs. These programs can help fund the deployment of SHS dryers in decentralized agro-processing units, thereby reducing post-harvest losses and increasing farmers' income.

7.2 Institutional Support Mechanisms

Research and development institutions are at the forefront of driving technological innovation and adaptation. In India, organizations like the Central Food Technological Research Institute (CFTRI), Indian Council of Agricultural Research (ICAR), and various State Agricultural Universities (SAUs) have made significant contributions in developing SHS drying prototypes, optimizing drying parameters, and conducting pilot-scale studies. These institutions often collaborate with industry partners to transfer lab-scale innovations to commercial applications.

In the global context, institutions like the United States Department of Agriculture-Agricultural Research Service (USDA-ARS) and Germany's Fraunhofer Institute for Interfacial Engineering and Biotechnology (IGB) have been pivotal in conducting interdisciplinary research on SHS drying. Their work not only provides scientific validation but also helps establish standards and best practices for the food processing industry.

Academic and technical institutions also play a vital role by incorporating SHS technology into their curricula, providing hands-on training, and conducting techno-economic feasibility analyses. Through initiatives like the National Agricultural Higher Education Project (NAHEP), educational institutions in India are encouraged to foster innovation and entrepreneurial skills among students, further supporting the adoption of sustainable technologies.

Financial and entrepreneurial support mechanisms are equally crucial. Development banks such as NABARD and SIDBI offer soft loans, capital subsidies, and credit-linked assistance for adopting clean processing technologies. Moreover, international organizations like the United Nations Industrial Development Organization (UNIDO) and the Global Environment Facility (GEF) provide funding, technical assistance, and capacity building for sustainable industrial practices, including SHS-based drying systems (Table 7.1).

Table 7.1 Global institutional framework for sustainable SHS drying.

Organization	Key role/function	Reference/initiative
FAO (UN Food & Agriculture Organization)	Promotes sustainable post-harvest management and energy-smart food systems.	FAO, *Energy-Smart Food for People and Climate* (2021)
UNDP (United Nations Development Programme)	Supports renewable energy for climate-resilient agri-value chains.	UNDP, *Climate Promise Projects* (2023)
UNIDO (United Nations Industrial Development Organization)	Promotes clean energy in agro-industries and solar drying tech.	UNIDO, *Sustainable Energy for Productive Use* (2022)
IRENA (International Renewable Energy Agency)	Facilitates global investment and data on solar drying systems.	IRENA, *Renewable Energy in Agri-Food Systems* (2021)
ISA (International Solar Alliance)	Promotes solar applications for agriculture and food drying.	ISA, *Scaling Solar Applications for Agriculture (SSA-A)* (2022)
World Bank/GCF	Provides finance for renewable-based agricultural technologies.	GCF, *Low Emission Climate Resilient Agriculture* (2023)
ISO/ Codex Alimentarius	Sets international standards for solar dryers and dried food safety.	ISO 9806; Codex, *Code of Practice for Dried Foods* (2023)

7.3 Challenges in Implementation

Despite its benefits, the widespread adoption of SHS drying faces several challenges. High initial capital investment is a major barrier, particularly for small and medium-sized enterprises (SMEs). SHS systems require specialized components and sophisticated control mechanisms, which increase upfront costs. Furthermore, a general lack of awareness about the technology, its benefits, and operational procedures limits its acceptance among food processors.

Another significant challenge is the limited availability of commercial-scale SHS systems and the lack of after-sales support. Many manufacturers are hesitant to enter this niche market due to uncertainty about demand and profitability. This creates a vicious cycle where the lack of supply discourages adoption, and low adoption rates dissuade manufacturers from investing in production. Policy implementation gaps also hinder progress. While numerous schemes support energy efficiency and rural processing, the absence of targeted incentives or specific inclusion of SHS technology in these programs restricts their effectiveness. There is also a need for technical standardization and certification protocols to ensure safety and performance consistency across different models and applications.

7.4 Future Perspectives for Sustainable SHS Drying

The future of SHS drying is promising, especially with ongoing advancements in automation, sensor technology, and energy integration. Incorporating artificial intelligence (AI) and the Internet of Things (IoT) into SHS systems can enable real-time monitoring and adaptive control of drying parameters, improving efficiency and product quality. Hybrid models that combine SHS with other technologies such as infrared, microwave, or solar drying can further enhance energy savings and operational flexibility.

Policy frameworks must evolve to keep pace with technological innovations. Future policies should explicitly recognize SHS drying as a green technology and provide dedicated incentives for its adoption. Establishing SHS technology clusters or industrial parks can facilitate shared infrastructure, reduce costs, and foster collaboration among stakeholders. Mandatory energy audits and technological upgrading mandates in food industries can also create a conducive environment for SHS deployment. International collaboration and knowledge sharing are essential to accelerate the adoption of SHS drying. Platforms such as the United Nations Climate Change Conferences (e.g., COP28), the World Food Summit, and the Global Forum on Food Security and Nutrition can serve as forums for exchanging best practices and aligning efforts. Collaborative research programs, technology transfer agreements, and joint funding initiatives can bridge the gap between developed and developing countries in terms of technological access and capability.

From a sustainability perspective, SHS drying aligns with several United Nations Sustainable Development Goals (SDGs), including SDG 7 (Affordable and Clean Energy), SDG 9 (Industry, Innovation and Infrastructure), and SDG 12 (Responsible Consumption and Production). Technology contributes to a circular economy by enabling waste heat recovery, reducing food waste through better preservation, and minimizing the carbon footprint of processing operations.

7.5 Conclusion

Superheated steam drying offers a viable solution to many of the challenges facing modern food processing, particularly in the context of sustainability and energy efficiency. However, its full potential can only be realized through a synergistic approach involving supportive policies, robust institutional mechanisms, and a strategic vision for future growth. Governments, academia, industry, and international organizations must collaborate to create an enabling environment that promotes research, innovation, and commercial adoption of SHS drying technologies. With the right mix of policy support and institutional commitment, SHS drying can become a cornerstone of the green food processing revolution.

References

Codex Alimentarius Commission. (2023). Code of practice for dried food.

Food and Agriculture Organization of the United Nations. (2021). Energy-smart food for people and climate. FAO.

Green Climate Fund. (2023). Low-emission climate-resilient agriculture. GCF.

International Organization for Standardization. (2013). ISO 9806:2013 — Solar energy — Solar thermal collectors — Test methods. ISO.

International Renewable Energy Agency. (2021). Renewable energy in agri-food systems. IRENA.

International Solar Alliance. (2022). Scaling solar applications for agriculture (SSA-A). ISA.

Kumar, P., Singh, R., & Kaur, M. (2023). "Superheated steam drying: technological advances and prospects." *Journal of Food Engineering*, 354, 111–123.

United Nations Development Programme. (2023). Climate Promise projects. UNDP.

United Nations Industrial Development Organization. (2022). Sustainable energy for productive use in agro-industries. UNIDO.

Glossary

Adsorption: The process by which molecules adhere to a solid or liquid surface; influences moisture migration during drying.

Automation: Use of control systems (PLCs, sensors, actuators) to operate drying equipment with minimal human intervention.

Biological material: Organic matter derived from plants, animals, or microbes used in food, agricultural, or bioprocessing applications.

Closed-loop system: A drying system where steam is recovered, dehumidified, reheated, and reused, minimizing energy loss and emissions.

Cogeneration (CHP): Combined heat and power generation, producing electricity and heat simultaneously to enhance energy efficiency.

Condensation: Conversion of steam or vapor into liquid water through cooling, often used for latent heat recovery.

Convective heat transfer: Transfer of heat through the motion of fluids such as air or steam during the drying process.

Digital twin: A virtual model of a physical drying system used for simulation, optimization, and predictive maintenance.

Drying kinetics: Rate and mechanism of moisture removal from materials during the drying process.

Energy efficiency: Measure of how effectively input energy is utilized for drying, minimizing losses through heat recovery or recycling.

Enthalpy: Thermodynamic property representing the total heat content of a system (internal energy plus flow work).

Evaporation: Conversion of liquid water into vapor during drying.

Fluidized bed dryer: Dryer where particles are suspended in a moving gas or steam stream, improving heat and mass transfer.

Fuzzy logic control: A smart control strategy that uses approximate reasoning to adjust drying parameters dynamically.

Heat and mass transfer: Simultaneous movement of heat into the material and moisture out of it during drying.

Hybrid drying: Combination of two or more drying methods (e.g., SHS + Microwave, SHS + Infrared) to improve efficiency and quality.

Infrared (IR) drying: Drying method using radiant energy to heat the product surface directly, accelerating evaporation.

Internet of Things (IoT): A network of interconnected sensors and devices that collect and share real-time data to optimize drying operations.

Latent heat: Amount of heat absorbed or released during a phase change without a temperature change.

Mechanical vapor recompression (MVR): System that compresses and reheats vapor for reuse, recovering energy and enhancing process sustainability.

Microwave (MW) drying: Technique using electromagnetic waves to generate volumetric heating, enabling rapid internal moisture removal.

Model predictive control (MPC): An advanced control technique that predicts future system behavior and adjusts parameters for optimal operation.

Moisture content: Amount of water present in a material, expressed as a percentage of total weight or dry basis.

Oxidation: Chemical reaction with oxygen that can deteriorate food color, flavor, and nutrients; prevented in SHS systems.

PLC (programmable logic controller): An industrial computer used to monitor and control automated drying processes.

Porosity: Measure of the void spaces within a material; affects drying rate and rehydration capacity.

Rehydration ratio: Indicator of how well a dried product regains its original properties after soaking in water.

Saturated steam: Steam in equilibrium with liquid water at a given temperature and pressure.

Shrinkage: Reduction in volume of a material due to moisture loss during drying.

Smart control system: AI- or algorithm-based control that optimizes drying parameters using real-time feedback.

Specific energy consumption (SEC): Energy required to remove a unit mass of water from a material during drying.

Steam generator: Device or boiler that produces steam by heating water for use in SSD systems.

Superheated steam (SHS): Steam heated above its saturation temperature, free from moisture droplets, used as a drying medium.

Superheated steam drying (SSD): A drying process using SHS in an oxygen-free environment, enhancing energy efficiency and product quality.

Thermal conductivity: Property determining how well a material conducts heat during drying.

Thermal efficiency: Percentage of total energy input effectively utilized in the drying process.

Thermodynamics: Branch of science concerned with energy transformations and heat transfer in physical systems.

Vacuum drying: Drying conducted under reduced pressure to lower the boiling point of water and protect heat-sensitive products.

Volatile organic compounds (VOCs): Evaporative organic compounds released during drying; minimized in closed-loop SHS systems.

Index

For Product Safety Concerns and Information please contact our EU representative GPSR@taylorandfrancis.com Taylor & Francis Verlag GmbH, Kaufingerstraße 24, 80331 München, Germany